~A BINGO BOOK~

Ecology
Bingo Book

COMPLETE BINGO GAME IN A BOOK

Written By Rebecca Stark

Educational Books 'n' Bingo

ISBN 978-0-87386-448-0

Educational Books 'n' Bingo

Printed in the U.S.A.

ECOLOGY BINGO DIRECTIONS

INCLUDED:

List of Terms

Templates for Additional Terms and Clues

2 Clues per Term

30 Unique Bingo Cards

Markers

1. **Either cut apart the book or make copies of ALL the sheets. You might want to make an extra copy of the clue sheets to use for introduction and review. Keep the sheets in an envelope for easy reuse.**

2. Cut apart the call cards with terms and clues.

3. Pass out one bingo card per student. There are enough for a class of 30.

4. Pass out markers. You may cut apart the markers included in this book or use any other small items of your choice.

5. Decide whether or not you will require the entire card to be filled. Requiring the entire card to be filled provides a better review. However, if you have a short time to fill, you may prefer to have them do the just the border or some other format. Tell the class before you begin what is required.

6. There are 50 terms. Read the list before you begin. If there are any terms that have not been covered in class, you may want to read to the students the term and clues before you begin.

7. There is a blank space in the middle of each card. You can instruct the students to use it as a free space or you can write in answers to cover terms not included. Of course, in this case you would create your own clues. (Templates provided.)

8. Shuffle the cards and place them in a pile. Two or three clues are provided for each term. If you plan to play the game with the same group more than once, you might want to choose a different clue for each game. If not, you may choose to use more than one clue.

9. Be sure to keep the cards you have used for the present game in a separate pile. When a student calls, "Bingo," he or she will have to verify that the correct answers are on his or her card AND that the markers were placed in response to the proper questions. Pull out the cards that are on the student's card keeping them in the order they were used in the game. Read each clue as it was given and ask the student to identify the correct answer from his or her card.

10. If the student has the correct answers on the card AND has shown that they were marked in response to the *correct questions,* then that student is the winner and the game is over. If the student does not have the correct answers on the card OR he or she marked the answers in response to *the wrong questions,* then the game continues until there is a proper winner.

11. If you want to play again, reshuffle the cards and begin again.

Have fun!

TOPICS INCLUDED

ABIOTIC FACTORS	FRESHWATER BIOMES
ACID RAIN	GLOBAL WARMING
ADAPTATION	GRASSLANDS
BIOMES	GREENHOUSE EFFECT
BIOTIC FACTORS	HABITAT
CARNIVORE	HAZARDOUS WASTE
COMMENSALISM	HERBIVORE
CONSERVATION	MARINE BIOMES
CONSUMER	NICHE
CORAL REEF	NITROGEN CYCLE
DECOMPOSERS	OCEANS
DEFORESTATION	OZONE LAYER
DESERTS	PARASITE
ECOLOGY	PHOTOSYNTHESIS
ECOSYSTEM	POLLUTION
ENDANGERED SPECIES	PREDATORS
ENERGY	PRIMARY PRODUCERS
EROSION	RECYCLING
ESTUARIES	SPECIES
EVOLUTION	SUCCESSION
EXTINCT	SYMBIOSIS
FOOD CHAIN	TROPHIC LEVELS
FOOD WEB	TROPICAL RAIN FOREST
FOREST	TUNDRA
FOSSIL FUELS	WETLANDS

Ecology Bingo

Additional Te

Choose as many additional topics as you would like and write them in the squares. Repeat each as desired.
Cut out the squares and randomly distribute them to the class.
Instruct the students to place their square on the center space of their card.

Ecology Bingo

Clues for Additional Terms

Write three clues for each of your additional terms.

<table>
<tr><td>

_____

1.

2.

3.

</td><td>

_____

1.

2.

3.

</td></tr>
<tr><td>

_____

1.

2.

3.

</td><td>

_____

1.

2.

3.

</td></tr>
<tr><td>

_____

1.

2.

3.

</td><td>

_____

1.

2.

3.

</td></tr>
</table>

© **Barbara M. Peller**

ABIOTIC FACTORS 1. These are the non-living components of a biosphere. 2. They include geological factors, such as rocks and minerals. 3. They include physical factors, such as temperature and precipitation.	**ACID RAIN** 1. It is precipitation heavy with nitric and sulfuric acid caused mostly by sulfur dioxide and nitrogen dioxide. 2. With a pH of less than 5.6, it causes fish and plant deaths. 3. This form of precipitation causes corrosion, groundwater pollution, and soil erosion.
ADAPTATION 1. It is the process by which individuals, populations, or species change in order to better survive under given environmental conditions. 2. It is the evolutionary process by which an organism becomes fitted to its environment. 3. It may refer to a structure or habit fitted for a particular environment or activity.	**BIOMES** 1. They are large climactically and geographically defined areas of ecologically similar communities. 2. These are classified according to the predominant vegetation and characterized by adaptations of organisms to that particular environment. 3. There are six major ones: Freshwater, Marine, Desert, Forest, Grassland and Tundra.
BIOTIC FACTORS 1. They are the living parts of the ecosystem with which an organism must interact. 2. These include everything that results from the activities of living things that directly or indirectly affect an organism in its environment. 3. Examples would be an organism's prey and predators.	**CARNIVORE** 1. It is a predatory, flesh-eating animal. It is the opposite of "herbivore." 2. It eats mostly herbivores. 3. Examples of this kind of animal that eats other animals include eagles, tigers, alligators and most humans.
COMMENSALISM 1. It is a relationship between two organisms where one benefits and the other is neither significantly harmed nor helped. 2. An example of this type of relationship would be a bird living in a tree without harming it. 3. An example of this type of relationship is mites who travel on the backs of beetles without harming them.	**CONSERVATION** 1. It is the planned management of a natural resource. 2. It is management of our use of the biosphere so that it yields the greatest sustainable benefit. 3. It includes the preservation, maintenance, sustainable use and restoration of the environment.
CONSUMER 1. Organisms at this trophic level eat autotrophs and herbivores. 2. A primary one gets its energy by eating producers, or autotrophs. 3. A secondary one gets its energy from eating autotrophs and herbivores.	**CORAL REEF** 1. This type of structure found in warm, shallow marine waters is formed by the stony skeletons of living organisms. 2. The Great Barrier Reef off the coast of Australia is the world's largest. 3. It is formed from coral polyps, which secrete an exoskeleton of calcium carbonate and live in colonies.

Ecology Bingo

DECOMPOSERS	DEFORESTATION
1. These organisms break down dead plants and animals. 2. When an organism dies, these organisms break down the organic matter into smaller particles and soluble forms to be used by primary producers. 3. Bacteria and fungi are important because as ____ they return nutrients to the soil.	1. This is the permanent destruction of indigenous forests and woodlands. 2. This is a growing concern because about 12 million hectares of forests are cleared annually. 3. One reason for concern is that the clearing of forests leads to an increase in the atmospheric CO_2 concentration.
DESERTS	**ECOLOGY**
1. These arid areas with sparse vegetation cover about one fifth of the Earth's surface. 2. They can be hot or cold. The Sahara is the largest hot one. 3. There is a great range in daily temperature in these areas. There is little moisture in the air to block the sun's rays, but when the sun goes down, the heat escapes.	1. It is the study of interrelationships between organisms and their environment. 2. A scientist who specializes in this area of study is called an ecologist. 3. This science is usually considered a branch of biology, but it is interdisciplinary and involves other branches of science as well.
ECOSYSTEM	**ENDANGERED SPECIES**
1. This unit comprises all the plants, animals, micro-organisms and non-living physical factors of an environment functioning as a unit. 2. It includes both the biotic and abiotic factors of an environment working as a unit. 3. Although similar to a biome, an ____ is much smaller. A biome comprises the many similar ____s found around the world.	1. It is the population of an organism at risk of becoming extinct. 2. The American peregrine falcon was on this list from 1970 to 1999. It was taken off because it recovered. The Mariana mallard was on the list from 1977 to 2004, when it became extinct. 3. The right whale, the ocelot and the leatherback sea turtle are examples.
ENERGY	**EROSION**
1. There are many forms of this, but they can all be put into two categories: kinetic and potential. 2. Some forms include heat, light, sound, electrical, nuclear and motion. 3. Sources may be categorized as renewable, such as wind, and nonrenewable, such as oil.	1. It is the carrying away, or displacement, of solids, usually by wind, water, ice or waves. 2. This natural process is often increased by poor human land use, such as deforestation and overgrazing. 3. Improved land-use practices can limit this process. Terrace-building and the planting of trees are two examples of what can be done.
ESTUARIES	**EVOLUTION**
1. These are places where freshwater rivers and streams flow into the ocean, mixing with the seawater. 2. Examples are bays, lagoons, harbors, inlets and sounds. 3. They are areas where fresh and marine waters mix.	1. It is the process of change by which new species develop from preexisting species over time. 2. Charles Darwin, a 19th-century naturalist, is considered the "Father of ____." 3. Darwin defined it as "descent with modification."

Ecology Bingo

EXTINCT	FOOD CHAIN
1. This term describes a species or a population that no longer exists. 2. Dinosaurs became this about 65 million years ago. 3. Dinosaurs, the dodo bird and the Tasmanian wolf are examples of animals who are now this.	1. This describes the path of energy from one organism to the next within an ecosystem. 2. In a marine one, plankton, microscopic plant life that floats in the ocean, is at the first level of this. 3. It starts with a primary producer and ends with a carnivore.
FOOD WEB	**FOREST**
1. It shows how several plants and animals are interconnected by different feeding relationships in a community. 2. Unlike a food chain, it branches out in all directions; a food chain follows a single path. 3. This form expands the food chain from a simple linear path to a complex network of interactions.	1. There are 3 main types of this biome: Tropical; Temperate; and Boreal, or Taiga. They are classified according to their latitude. 2. A coniferous ___ is made up mostly of cone-bearing, or coniferous, trees. Examples are firs, hemlocks, pines and spruces. 3. Trees in a temperate deciduous ___ lose their leaves in fall and regrow them in the spring.
FOSSIL FUELS	**FRESHWATER BIOMES**
1. They include coal, oil and natural gas. They are nonrenewable resources because they take millions of years to form. 2. Reserves of these resources are being depleted much faster than new ones are being formed. 3 These nonrenewable fuels provide more than 85% of all the energy consumed in the U.S.	1. These biomes includes three types of regions: ponds and lakes, streams and rivers, and wetlands. 2. Plants and animals in these biomes would not be able to survive in areas of high salt concentration. 3. This low-saline, aquatic biome covers one fifth of the Earth's surface.
GLOBAL WARMING	**GRASSLANDS**
1. It is the increase in the average temperature of the Earth's near-surface air and oceans. **2.** Most scientists working on climate change agree with the Intergovernmental Panel on Climate Change that ___ is occurring. 3. One of the reasons for ___ is the greenhouse effect.	1. These are large areas with rolling hills of grasses and wildflowers. 2. They are found in areas where there is too little precipitation to support the growth of a forest, but not enough to prevent the formation of a desert. 3. Three types of ___ are steppe, prairie and savanna.
GREENHOUSE EFFECT	**HABITAT**
1. ___ is the ability of the atmosphere to trap and recycle energy emitted by the Earth surface. 2. The main gas contributing to human-caused enhancement of this process is carbon dioxide. Others include methane, chlorofluorocarbons and nitrous oxide. 3. Scientists believe an increase in the major gases responsible for this will lead to global warming.	1. It is the natural environment in which a particular species lives. 2. It is the place where a plant and/or animal population lives and its surroundings, both living and nonliving. 3. It is the physical and biological environment on which a particular species depends for its survival.

Ecology Bingo

HAZARDOUS WASTE	HERBIVORE
1. This term refers to toxic chemicals, radioactive materials, and biologic or infectious materials. 2. Most of it results from industrial processes that yield unwanted by-products, defective products and spilled materials. 3. The Environmental Protection Agency (EPA) protects us from the dangers of this by setting and enforcing national standards for its disposal.	1. This kind of animal feeds on plants. 2. It is a primary consumer. 3. This kind of animal needs more energy than a carnivore. Many, like cows and sheep, eat all day long.
MARINE BIOMES	**NICHE**
1. Included in these biomes are coral reefs, estuaries and oceans. 2. Algae from these biomes supply much of the Earth's oxygen supply and take in a huge amount of atmospheric carbon dioxide. 3. Unlike freshwater biomes, these involve a medium to high salinity.	1. It is the ecological role of a species or population in a community. 2. It is the function of a species in an ecosystem. 3. According to the principle of competitive exclusion, if two species compete for the same ___, one will eventually displace the other.
NITROGEN CYCLE	**OCEANS**
1. Its 4 main processes include nitrogen fixation, decay, nitrification by bacteria and denitrification. 2. It is the cycling of nitrogen from the air & soil to plants & animals and back to the environment. 3. Its first 3 processes remove nitrogen from the atmosphere and pass it through ecosystems. The 4th reduces nitrates to nitrogen gas, replenishing the atmosphere.	1. These are the largest ecosystems in the world. 2. These ecosystems are divided into 4 zones: intertidal, pelagic, abyssal, and benthic. 3. Evaporation of seawater from these large bodies of water provide rainwater for the land.
OZONE LAYER	**PARASITE**
1. It is the layer in the stratosphere—about 10 to 20 miles above Earth's surface—where most atmospheric O_3 is concentrated. 2. It acts as a shield that protects life on Earth from the sun's harmful radiation. 3. This is being depleted because of chlorofluorocarbons and other chlorine- and bromine-producing substances.	1. It is an organism that lives in a close relationship with another organism, causing it harm. 2. It is dependent upon its host to survive. 3. It benefits, but its host is harmed in some way.
PHOTOSYNTHESIS	**POLLUTION**
1. This is the process by which green plants use water and carbon dioxide to create glucose and oxygen. 2. This process takes energy from the sun and converts it into a storable form which plants use for their own life processes. 3. Animals provide the carbon dioxide needed for this process and get oxygen in return.	1. In general, it is the introduction of contaminants into an environment. 2. Burning fossil fuels and emissions from automobiles are causes of air ___. 3. In 1978 the groundwater ___ at Love Canal was traced to organic solvents and dioxins from an industrial landfill.

Ecology Bingo

PREDATORS	PRIMARY PRODUCERS
1. It is a creature that captures another as food. 2. The creature that is captured by this creature is called prey. 3. These secondary consumers are also called carnivores.	1. They bring energy into an ecosystem from inorganic sources. All life directly or indirectly relies on them. 2. Another name for organisms at this trophic level is "autotrophs." 3. Photosynthetic green plants and others at this trophic level form the base of any food chain.
RECYCLING	**SPECIES**
1. It is the reprocessing of old materials into new products to prevent waste. 2. Some products commonly included in this practice are glass, paper, metal and plastics. 3. This practice is the third component environmentalists refer to as the "3 R's." The other two are "Reduce" and "Reuse."	1. Each ___ is given a binomial name consisting of the generic name, or genus, and a specific descriptive term. 2. With some exceptions, it comprises related organisms or populations potentially capable of interbreeding and producing fertile offspring. 3. Humans are the only ___ to have the binomial name *Homo sapiens*.
SUCCESSION	**SYMBIOSIS**
1. It is the gradual supplanting of one community of plants by another. 2. It is the progressive transformation or origination of an ecological community as new plant and animal species come into an area. 3. If it begins where no soil is initially present, this change in the ecological community is called primary; otherwise, it is called secondary.	1. It is a close ecological relationship between 2 or more individuals of different species. 2. Mutualism, in which both species benefit; commensalism, in which 1 benefits and the other is unaffected; and parasitism, in which 1 benefits and the other is harmed are examples. 3. Pollination is an example of mutualism, a form of ___ in which individuals of both species benefit.
TROPHIC LEVELS	**TROPICAL RAINFORESTS**
1. They are the levels of a food chain. Some energy and/or biomass is lost at each level. 2. At the first level are the primary producers. Next are the primary consumers, then the secondary and sometimes tertiary consumers. 3. They describe the position an organism occupies in a food chain. Green plants form the first one.	1. They have tall trees, a warm climate, and receive at least 80 inches of rain a year. Some get more than one inch every day! 2. They are vertically divided into at least five layers: the overstory, the canopy, the understory, the shrub layer, and the forest floor. Most plants and animals live in the canopy. 3. The Amazon ___ is the largest in the world.
TUNDRA	**WETLANDS**
1. It is the coldest biome. 2. The Alpine ___ is located at very high altitudes where trees cannot grow. 3. The Arctic ___ encircles the north pole and extends south to the coniferous forests of the taiga.	1. These are areas of land consisting of soil that is saturated with moisture and can support vegetation adapted for life in saturated soil. 2. Swamps, marshes and bogs are examples. 3. Saturation with water is the dominant factor determining the nature of the soil and the types of plant and animal communities living in these areas.

Ecology Bingo

Ecology Bingo

Endangered Species	Acid Rain	Commensalism	Food Web	Freshwater Biomes
Fossil Fuels	Adaptation	Succession	Herbivore	Trophic Levels
Biomes	Wetlands		Marine Biomes	Ecology
Recycling	Deserts	Tropical Rainforest	Grasslands	Nitrogen Cycle
Ozone Layer	Estuaries	Global Warming	Symbiosis	Primary Producers

Ecology Bingo: Card No. 1

Ecology Bingo

Recycling	Biomes	Hazardous Waste	Tundra	Food Chain
Nitrogen Cycle	Decomposers	Biotic Factors	Deserts	Greenhouse Effect
Coral Reef	Estuaries		Erosion	Tropical Rainforest
Photosynthesis	Oceans	Wetlands	Pollution	Primary Producers
Trophic Levels	Succession	Global Warming	Fossil Fuels	Symbiosis

Ecology Bingo

Recycling	Tropical Rainforest	Decomposers	Grasslands	Biomes
Herbivore	Adaptation	Consumer	Acid Rain	Forest
Deserts	Succession		Greenhouse Effect	Carnivore
Wetlands	Coral Reef	Ozone Layer	Photosynthesis	Hazardous Waste
Symbiosis	Fossil Fuels	Global Warming	Pollution	Food Chain

Ecology Bingo

Wetlands	Greenhouse Effect	Commensalism	Fossil Fuels	Food Chain
Habitat	Conservation	Acid Rain	Tundra	Biomes
Marine Biomes	Photosynthesis		Freshwater Biomes	Food Web
Tropical Rainforest	Ecosystem	Succession	Global Warming	Biotic Factors
Abiotic Factors	Trophic Levels	Oceans	Symbiosis	Ecology

Ecology Bingo

Trophic Levels	Freshwater Biomes	Deserts	Biotic Factors	Fossil Fuels
Habitat	Tropical Rainforest	Consumer	Erosion	Adaptation
Commensalism	Ecology		Herbivore	Evolution
Primary Producers	Food Chain	Endangered Species	Pollution	Deforestation
Decomposers	Global Warming	Biomes	Wetlands	Marine Biomes

Ecology Bingo

Carnivore	Greenhouse Effect	Hazardous Waste	Food Chain	Ecology
Grasslands	Deserts	Deforestation	Acid Rain	Biomes
Tundra	Abiotic Factors		Conservation	Erosion
Global Warming	Ozone Layer	Pollution	Oceans	Commensalism
Nitrogen Cycle	Biotic Factors	Endangered Species	Marine Biomes	Ecosystem

Ecology Bingo

Endangered Species	Greenhouse Effect	Evolution	Herbivore	Decomposers
Nitrogen Cycle	Food Chain	Estuaries	Adaptation	Habitat
Hazardous Waste	Food Web		Erosion	Conservation
Wetlands	Photosynthesis	Consumer	Recycling	Coral Reef
Global Warming	Fossil Fuels	Pollution	Oceans	Carnivore

Ecology Bingo

Marine Biomes	Greenhouse Effect	Energy	Grasslands	Conservation
Habitat	Commensalism	Tundra	Ecology	Biotic Factors
Ecosystem	Parasite		Food Chain	Freshwater Biomes
Symbiosis	Wetlands	Recycling	Abiotic Factors	Photosynthesis
Succession	Global Warming	Oceans	Deserts	Nitrogen Cycle

Ecology Bingo

Erosion	Decomposers	Estuaries	Ecosystem	Fossil Fuels
Abiotic Factors	Food Chain	Marine Biomes	Deserts	Greenhouse Effect
Forest	Endangered Species		Adaptation	Energy
Deforestation	Primary Producers	Ozone Layer	Herbivore	Evolution
Photosynthesis	Pollution	Consumer	Recycling	Freshwater Biomes

Ecology Bingo: Card No. 9

© Barbara M. Peller

Ecology Bingo

Recycling	Grasslands	Conservation	Tundra	Ecosystem
Ecology	Biotic Factors	Acid Rain	Adaptation	Food Chain
Parasite	Greenhouse Effect		Food Web	Coral Reef
Ozone Layer	Primary Producers	Deforestation	Pollution	Forest
Consumer	Nitrogen Cycle	Hazardous Waste	Trophic Levels	Marine Biomes

Ecology Bingo

Carnivore	Greenhouse Effect	Deserts	Deforestation	Nitrogen Cycle
Energy	Forest	Herbivore	Erosion	Acid Rain
Habitat	Food Chain		Hazardous Waste	Estuaries
Consumer	Biomes	Pollution	Fossil Fuels	Recycling
Abiotic Factors	Global Warming	Endangered Species	Oceans	Decomposers

Ecology Bingo

Decomposers	Freshwater Biomes	Forest	Grasslands	Erosion
Estuaries	Nitrogen Cycle	Commensalism	Oceans	Adaptation
Endangered Species	Evolution		Ecology	Tundra
Global Warming	Photosynthesis	Food Chain	Recycling	Habitat
Greenhouse Effect	Energy	Parasite	Abiotic Factors	Biotic Factors

Ecology Bingo

Deforestation	Freshwater Biomes	Carnivore	Forest	Ecology
Commensalism	Energy	Food Chain	Erosion	Coral Reef
Grasslands	Decomposers		Estuaries	Evolution
Marine Biomes	Pollution	Conservation	Parasite	Recycling
Global Warming	Primary Producers	Oceans	Endangered Species	Herbivore

Ecology Bingo

Fossil Fuels	Food Chain	Deserts	Erosion	Abiotic Factors
Biotic Factors	Endangered Species	Forest	Adaptation	Greenhouse Effect
Deforestation	Food Web		Hazardous Waste	Consumer
Primary Producers	Pollution	Parasite	Conservation	Carnivore
Global Warming	Tundra	Coral Reef	Nitrogen Cycle	Marine Biomes

Ecology Bingo

Herbivore	Erosion	Deserts	Decomposers	Grasslands
Carnivore	Hazardous Waste	Acid Rain	Commensalism	Abiotic Factors
Ecology	Endangered Species		Biomes	Greenhouse Effect
Global Warming	Forest	Energy	Pollution	Deforestation
Nitrogen Cycle	Photosynthesis	Oceans	Ecosystem	Estuaries

Ecology Bingo: Card No. 15

Ecology Bingo

Conservation	Forest	Energy	Ecosystem	Predators
Tundra	Coral Reef	Evolution	Habitat	Food Web
Deforestation	Freshwater Biomes		Ecology	Estuaries
Wetlands	Biotic Factors	Global Warming	Niche	Recycling
Abiotic Factors	Species	Oceans	Photosynthesis	Greenhouse Effect

Ecology Bingo

Consumer	Niche	Extinct	Forest	Fossil Fuels
Herbivore	Abiotic Factors	Pollution	Food Web	Evolution
Erosion	Marine Biomes		Species	Energy
Primary Producers	Nitrogen Cycle	Recycling	Deserts	Coral Reef
Ozone Layer	Deforestation	Decomposers	Grasslands	Freshwater Biomes

Ecology Bingo: Card No. 17

Ecology Bingo

Ecosystem	Parasite	Biotic Factors	Deforestation	Tundra
Greenhouse Effect	Consumer	Ozone Layer	Ecology	Abiotic Factors
Erosion	Coral Reef		Extinct	Commensalism
Primary Producers	Acid Rain	Pollution	Recycling	Hazardous Waste
Species	Forest	Deserts	Niche	Carnivore

Ecology Bingo

Ecology	Carnivore	Forest	Energy	Ecology
Herbivore	Grasslands	Greenhouse Effect	Decomposers	Food Web
Niche	Fossil Fuels		Adaptation	Biomes
Hazardous Waste	Species	Ozone Layer	Photosynthesis	Extinct
Commensalism	Predators	Nitrogen Cycle	Marine Biomes	Oceans

Ecology Bingo

Parasite	Niche	Grasslands	Forest	Oceans
Biotic Factors	Estuaries	Habitat	Ozone Layer	Tundra
Freshwater Biomes	Evolution		Wetlands	Acid Rain
Trophic Levels	Succession	Symbiosis	Photosynthesis	Species
Tropical Rainforest	Marine Biomes	Predators	Recycling	Extinct

Ecology Bingo: Card No. 20

Ecology Bingo

Herbivore	Carnivore	Habitat	Forest	Trophic Levels
Freshwater Biomes	Extinct	Conservation	Energy	Endangered Species
Coral Reef	Nitrogen Cycle		Niche	Deserts
Ozone Layer	Decomposers	Species	Primary Producers	Marine Biomes
Wetlands	Predators	Oceans	Consumer	Photosynthesis

Ecology Bingo

Ecosystem	Hazardous Waste	Extinct	Commensalism	Deforestation
Tundra	Grasslands	Biomes	Energy	Adaptation
Biotic Factors	Food Web		Endangered Species	Evolution
Species	Primary Producers	Photosynthesis	Acid Rain	Fossil Fuels
Predators	Consumer	Niche	Coral Reef	Habitat

Ecology Bingo

Conservation	Niche	Decomposers	Commensalism	Oceans
Carnivore	Parasite	Nitrogen Cycle	Herbivore	Acid Rain
Hazardous Waste	Deforestation		Symbiosis	Endangered Species
Coral Reef	Predators	Species	Consumer	Photosynthesis
Trophic Levels	Succession	Marine Biomes	Ozone Layer	Extinct

Ecology Bingo

Conservation	Parasite	Fossil Fuels	Niche	Energy
Extinct	Oceans	Habitat	Tundra	Endangered Species
Evolution	Ecosystem		Deforestation	Coral Reef
Trophic Levels	Symbiosis	Species	Consumer	Freshwater Biomes
Tropical Rainforest	Wetlands	Predators	Grasslands	Succession

Ecology Bingo: Card No. 24

Ecology Bingo

Wetlands	Habitat	Niche	Deserts	Extinct
Acid Rain	Primary Producers	Herbivore	Conservation	Adaptation
Freshwater Biomes	Energy		Symbiosis	Species
Biomes	Trophic Levels	Succession	Predators	Food Web
Oceans	Fossil Fuels	Biotic Factors	Abiotic Factors	Tropical Rainforest

Ecology Bingo

Extinct	Niche	Hazardous Waste	Tundra	Ecosystem
Ozone Layer	Grasslands	Energy	Parasite	Conservation
Primary Producers	Symbiosis		Food Web	Wetlands
Consumer	Commensalism	Trophic Levels	Predators	Species
Evolution	Abiotic Factors	Deserts	Succession	Tropical Rainforest

Ecology Bingo

Hazardous Waste	Biotic Factors	Niche	Parasite	Estuaries
Trophic Levels	Symbiosis	Herbivore	Species	Adaptation
Pollution	Succession		Predators	Wetlands
Ecosystem	Carnivore	Habitat	Tropical Rainforest	Acid Rain
Abiotic Factors	Food Web	Extinct	Biomes	Evolution

Ecology Bingo

Ecology	Parasite	Biomes	Niche	Conservation
Estuaries	Extinct	Symbiosis	Tundra	Food Web
Succession	Coral Reef		Evolution	Ozone Layer
Recycling	Ecosystem	Nitrogen Cycle	Predators	Species
Commensalism	Erosion	Abiotic Factors	Tropical Rainforest	Trophic Levels

Ecology Bingo

Extinct	Parasite	Ecosystem	Herbivore	Erosion
Primary Producers	Ozone Layer	Habitat	Evolution	Biomes
Freshwater Biomes	Symbiosis		Adaptation	Niche
Estuaries	Trophic Levels	Food Chain	Predators	Species
Conservation	Energy	Tropical Rainforest	Carnivore	Succession

Ecology Bingo

Fossil Fuels	Niche	Tundra	Erosion	Species
Acid Rain	Parasite	Hazardous Waste	Food Web	Adaptation
Primary Producers	Deforestation		Evolution	Habitat
Tropical Rainforest	Carnivore	Commensalism	Predators	Symbiosis
Trophic Levels	Decomposers	Succession	Extinct	Biomes

Ecology Bingo: Card No. 30

www.ingramcontent.com/pod-product-compliance
Lightning Source LLC
Chambersburg PA
CBHW080721220326
41520CB00056B/7286